CANCER

HOROSCOPE

& ASTROLOGY

2021

Published by Mystic Cat Press

Suite SM-2380-6403

14601 North Bybee Lake Court

Portland, Oregon 97203

Phone: +1 (805) 308-6503

SiaSands@hotmail.com

Contents

Acknowledgment:

To my family, thank you for being there and accepting my wildness.

This book is dedicated to those with an open heart, an open mind, and a willingness to plumb the mysteries of life.

You make this world a better place.

CANCER 2021
Horoscope & Astrology

CANCER

Cancer Dates: June 21 to July 22
Symbol: Crab
Element: Water
Planet: Moon
House: Fourth
Colors: Silver, white

2021 CANCER OVERVIEW

2021 inspires and delights with three gorgeous Supermoons in the first half of the year. A new mission is coming soon, it revitalizes your spirit and brings a visionary quest you can embrace. It is a project that overhauls your situation and offers you room to grow your life. If you have been feeling frazzled recently, this is set to shift. Inspiring energy soon provides you with a welcome boost. News arrives, which brings a big venture. You've got big plans brewing, your creativity is rising to meet your vision. An innovative idea brings an inspiring path. You can plot a course forward and incorporate some of your more enormous ideas into a robust phase of growth. An area you begin, soon blooms, providing you with tangible feedback. It gives you the information needed to let go of the safety rails and dive deep into a new adventure. It does set the tone of unique and inspiring options to be revealed in 2021.

On February 12th, we ring in the Chinese New Year of the Ox, this is an important event, it brings grounded energy into your life. You may underestimate what you're capable of, you can increase your potential by exploring new avenues of learning and growth. You are highly creative, and you can harness this to your advantage and make progress towards achieving your vision.

Mercury Retrograde gets up to tricks in 2021, taking time to nurture your spirit transports healing energy to your mind, body, and spirit. If you have found that things have been chaotic, create space to ground and center your energy. Your core foundations will carry you through, time is everything, things will soon flow forward towards smoother waters.

THE ECLIPSES

LUNAR & SOLAR

Solar eclipses can only occur during a New Moon phase. This is when the Moon moves between Earth and the Sun, and these three celestial bodies form a straight line: Earth–Moon–Sun.

A Lunar eclipse occurs when the earth stands between the moon and the Sun, this obscures the light of the Sun from the moon. The moon herself has no light source of her own, as she simply reflects the light of the Sun. A lunar eclipse occurs during a Full Moon and usually marks endings, transitions, or other life cycle culmination points.

Any eclipse is a significant event in astrological circles, eclipses have fascinated scientists for centuries. Eclipses are dramatic tools that instigate change in your life. An eclipse is wild, free, expansive, and explosive, the wild cards of astrology, you never quite know what you get until it happens. An eclipse can uproot, surprise, inspire, motivate, and really become an active catalyst for change. Eclipses remove the shutters, they make you aware of areas that need to be changed and often spotlight an entirely new direction to explore. Eclipses inspire change and work rapidly to see forward motion occurring.

WORKING WITH THE MOON

2021 delights with three gorgeous Supermoon's. A supermoon is when the moon is at its closest approach to Earth, which occurs during a full or new moon. The effect on the ocean's tides is most significant when there is a full or new moon. This tidal force is concentrated during the super moon, it can cause the ocean tides to rise by an extra inch or two compared to a regular full moon. Super moons are they invite you to look at your life, to reveal areas which you usually keep hidden. High in the night sky, they illuminate a great deal of information should you choose to work with this sacred energy. Connecting with this information gives you a fantastic opportunity to expand your life, to reveal areas that are ready to be developed.

As the moon peaks, it naturally begins to wane, and as the moon heads towards the next gravitational peak, the new moon phase, it has a cleansing effect on your emotional awareness. This helps you remove from your life all the things that need to be released, the areas which limit progress said no real good while they are kept within your spirit. Heading into the new moon gives your excellent opportunity to connect with the mysterious darkness. It is a healing time that brings a powerful sense of cleansing. This removes the outworn energy and makes space for new opportunities to flow into your world as the moon fills once again into a full shining globe.

PLANETARY RETROGRADES

The Retrograde phase is when a planet appears, when observed from Earth, to reverse direction. This happens due to an optical illusion caused by differences in orbit. The retrograde motion can have a negative influence on your life. The planet Mercury is the best-known planet for retrograde phases. This is because Mercury is the fastest planet in our solar system, and it enters a retrograde motion between three to four times a year, for about three weeks at a time. Mercury is a planet that rules communication, so you can expect frequent misunderstandings, scheduling problems, and disagreements during a Mercury Retrograde phase. Here is a quick reference guide to the retrogrades in 2021.

MERCURY: 3 RETROGRADES IN 2021

VENUS: 1 RETROGRADE IN 2021

MARS: NO RETROGRADE IN 2021

JUPITER: 1 RETROGRADE IN 2021

SATURN: 1 RETROGRADE IN 2021

URANUS: 2 RETROGRADE IN 2021

NEPTUNE: 1 RETROGRADE IN 2021

PLUTO: 1 RETROGRADE IN 2021

NODE: 1 RETROGRADE IN 2021

LILITH: NO RETROGRADE IN 2021

CHIRON: 1 RETROGRADE IN 2021

2021

CANCER HOROSCOPE

Four Weeks Per Month

- Week 1 – Days 1 - 7
- Week 2 – Days 8 - 14
- Week 3 – Days 15 - 21
- Week 4 – Days 22 – Month-end

Time is set to Coordinated Universal Time Zone (UT±0)

JANUARY ASTROLOGY

January 3, 4 - Quadrantids Meteor Shower.

The Quadrantids meteor shower run yearly from January 1-5. The Quadrantids meteor shower peaks this year on the night of the 3rd and morning of the 4th.

January 6 – Last Quarter Moon in Libra.

This Moon phase occurs at 09.37 UTC.

January 13 – New Moon in Capricorn.

This new moon phase occurs at 05:02 UTC. This cleans the slate and brings a fresh start. This is an excellent time to view galaxies and stars as there is no moonlight to obscure your view of the universe.

January 20 – First Quarter Moon in Aries.

This Moon phase occurs at 21.02 UTC.

January 24 – Mercury at Greatest Eastern Elongation.

The planet Mercury reaches greatest eastern elongation of 18.6 degrees from the Sun. This occurs at 02.00 UTC. Look for Mercury low in the sky just after sunset.

January 28 - Full Moon in Leo.

This phase occurs at 19:16 UTC. Full Wolf Moon. It has also been known as the Old Moon and the Moon After Yule. The Full Moon illuminates and draws new options to light.

January 29 – Jupiter in Conjunction with the Sun.

The planet Jupiter in Conjunction with the Sun. This occurs at 01:00 UTC.

January 30 – Mercury Retrograde begins in Aquarius.

During a retrograde period, it isn't the right time to move forward in any practical venture. Be prepared for misunderstandings and miscommunications to be prevalent.

JANUARY WEEK ONE

The Quadrantids Meteor Shower blazes across the night sky this week. You reach a tipping point soon, which enables you to embrace a newfound flow of abundance. There is an area that you dream about, and you are setting the correct intentions to help manifest this goal in due course. It brings a carefree time where you explore what life has to offer. It does take the weight of responsibility from your shoulders, a new lease of life has you stepping out and enjoying a refreshing chapter that is highly social. It does indicate a time of heightened security and stability is ahead. You open a groundbreaking chapter of potential, it does have you plotting a course towards a grand vision. An area you focus on brings growth. It offers an option that advances your situation, a new sense of belonging blooms, and this gives you a vital sense of purpose. It kick-starts a bountiful time with refreshing options to explore. You can expect life to switch up a gear soon. Changes ahead bring a piece of positive news. A new sense of creativity brings innovative ideas, you discover a path that draws excitement and optimism. It does bring a venture you can sink your teeth into, your talents are ready to shine, it is an auspicious chapter to expand your horizons and head out to see the sights that tempt you forward. Indeed, clarity brings the moment you have been seeking. It does plant you on the ground that enables you to craft a vision that is in alignment with your highest calling. You reveal a unique path that offers a rich landscape to explore. Your imagination and intuition blend an excellent brew of intoxicating potential. You rediscover your sense of adventure and reawaken to the luxurious view of new possibilities.

The New Moon in Capricorn at the week's end brings a fork in the road. An unresolved matter from your past arises, this is actually positive news, it enables you to make peace with what has gone before. It brings a resolution that allows you to keep moving forward instead of dwelling on what can't be changed. When the limiting energy is released, you feel empowered and lighter. Attending to this unfinished business is a valuable tool towards healing the past. There is a shift that brings change, it releases deep-seated insecurities, it sees your confidence rising, and it does bring a new path to explore. Stepping out towards a new venture feels refreshing. It lights up a way of adventure, progression, and growth. It is a time of broadening your perception and going wide on a new adventure. If you have struggled to create lifestyle changes, this will no longer hold you back. Your pioneering style blazes a trail that offers exciting new options. Your potential is on the rise, it does take you further, and launches you toward a successful chapter. It is a spontaneous time where you may act on an impulse that speaks to your heart. Having the proper outlet for your passion draws a section of new adventures. You race forward towards an exciting landscape filled with your dreams for future possibilities. You're unwilling to settle for less, setting the bar higher does bring an elevated option into your world. You have learned the lessons of the past and now step into your own authority with the confidence required to obtain your vision. It does see life shines with new potential. A path you initiate has room to progress into something substantial. This relates to plotting a course towards a larger goal, one that inspires your heart.

This time speaks of communication, news, and information arriving. It does bring a shift that inspires your mind. There is the energy that comes together nicely, it's an especially appropriate time to open the door on a fresh chapter. It brings new options, and this enables you to develop a situation that inspires your mind. Setting your intentions does bring new goals into focus. It links you up with a chapter where you can plot a course towards new goals. Remarkably, it is a time that is freewheeling, limitations are released, you lay a solid foundation, and this gives you the confidence to take your situation to a new level. It does bring adventures and the sense of wildness that inspires your mind. It draws an attraction that is difficult to resist, dynamic news arrives that is the icing on the cake. You work hard to build a solid foundation, it does have the desired effect and draws harmony into your life. It brings the completion of a significant chapter, it encourages you to move in alignment with your heart. It depicts a path that brings greater security, you take steps necessary to manifest abundance. It does bring a landscape that offers room to progress your vision. You are given the green light to get the ball rolling on achieving a long-held dream. A pathway opens that enables you to chase your imagination and explore possibilities that inspire you to push back the barriers. It sweeps in the refreshing potential, it brings a busy time, there is a gateway ahead, crossing through this portal does bring a decisive shift towards advancing your situation. Veering out of your safety zone draws fruitful results.

Jupiter goes into Conjunction with the Sun the day after the Full Moon in Leo. Jupiter rules luck, growth, wisdom, and fortune. Life gets a boost from this cosmic alignment, it provides you with a good sense of intuition, you can spot the signs, and this kicks off a curious chapter that enables you to go after a substantial goal. There are changes ahead that inspire your mind, it refreshes your life, and it does see you departing dramatically from your everyday routine. It brings new options that land you in an exciting environment. There is a chance to celebrate a more social environment soon. It may see you involved with a community endeavor. Heightened options to mingle does keep you on your toes, it is a boost, and this sees opportunities to network with others who are on the same page. Life begins to take on a rosy hue, it shakes off any heaviness that has held you back. You move onward towards a brighter future. There's lots of activity coming that brings a focused goal orientated chapter. It does harness a sense of optimism; expanding your world, you embark on the adventurous time of developing a path that sparks with refreshing potential. You may decide to take a leap of faith and dive into this time of growth and good fortune. A lofty goal comes into focus. It is a time that sparks with new potential.

Mercury Retrograde begins in Aquarius at weeks end. You can navigate through this unsettling phase by keeping focused on the destination. Delay unnecessary deals, it's not the best time to sign any legal documents.

FEBRUARY ASTROLOGY

February 2 – Imbolc

Harness the element of fire to create something new. Inspiration, motivation, and creativity are rising. The earth is waking after winter's long sleep.

February 4 – Last Quarter Moon in Scorpio.

This Moon phase occurs at 17.37 UTC.

February 8 – Mercury at Inferior Conjunction.

The planet Mercury at Inferior Conjunction. This occurs at 14:00 UTC.

February 11 - New Moon in Aquarius.

This phase occurs at 19:06 UTC. This is an excellent time to view galaxies and stars as there is no moonlight to obscure your view of the universe. This is a time of rebirth and renewal. Create space for something new to arrive.

February 12 – Chinese New Year (Ox)

February 19 – First Quarter Moon in Taurus.

This Moon phase occurs at 18.47 UTC.

February 21 – Mercury Retrograde ends in Aquarius.

You can now move forward with any delayed plans that you have been putting off due to the Mercury Retrograde phase. Relationships should soon improve as miscommunications are overcome

February 27 - Full Moon in Virgo.

The Moon is on the opposite side of the Earth as the Sun and will be fully illuminated. This phase occurs at 08:17 UTC. This full moon is known as the Full Snow Moon. Powerful energy lights a path forward. You can attract and manifest excellent results during the complete moon phase.

There is a sense of coming home soon when you visit revisit a past area. It does create space to resolve outworn energy, it sees your life reach the full circle. Taking time to contemplate the path ahead sets the scene for a beautiful journey to unfold over the coming chapter. You align with your core dreams and vision. Something is coming that needs your full attention, doing healing work at this time nurtures the correct environment for this exciting potential. While a turn of events ahead can feel chaotic, it is for the best. You can transition towards a new area that sees your life lighting up substantially. As you explore new possibilities, you discover a bond that makes your heart sing. It brings a spectacular chapter that sparkles with good fortune. You deserve to chase your happiness, and life will support your vision. If you open to the opportunities which seek to emerge. Give your body the rest it deserves and let your spirit rebuild. If you been waiting for news, something will be revealed soon. It does bring a big sky dreaming time. It helps you come up with a solid plan that will be instrumental in dealing with any emotional blocks that hold you back. It illuminates the path forward and brings a new chapter into your life. There might be an area that has put the brakes on advancing your goals. Something is hindering your progress. Yet the news is imminent to provide you with illuminating information. It does clear the path and helps you gain the knowledge necessary to make strides at improving your situation. It does bring a powerful option that lets you head towards growth. This lines up your trajectory favorably, it inspires your mind to continue to chase your vision.

The New Moon this week does wipe the slate clean on many levels. Your life has undergone many changes, this can feel unsettling, but you have nailed the ability to make the most of the environment you find yourself in. Shuffling the deck of fate, you create an entirely new life path that resonates warmly. There is currently a window of opportunity that enables you to reflect and integrate recent events. It is a valuable time to take inventory of your goals, it's never too late to tweak or even change the destination you have in mind. A flash idea may arrive suddenly and sweep you off to a new adventure. Your creativity is going to take you places, it brings new options that bring out your optimistic side. Surprise news calls your name soon. You are on a path to revolutionize your world. It does see a transition occurring that transforms your trajectory and brings new goals into focus. A vision you set your sights on does offer room to grow. A message arrives that sparks new potential. It does bring a good situation that is perfect for your talents. As this area makes an appearance in your life, you kick into high gear and shift your focus towards progressing your situation forward. It does launch a new and industrious chapter, it supercharges your dreams and gives you plenty to focus on in the section ahead. You have worked hard to improve your life. You are ready to transition towards a path that reflects the care and dedication you give to your world. Moving in alignment with your vision does see you focusing your energy correctly, it lets you discard areas that do not resonate with your vision. Under this powerful influence, you can achieve an active phase of advancing your dreams.

Mercury Retrograde ends at weeks end. Life gives you a fantastic opportunity to utilize your creative side, and fan the flames of potential. A moment of transformation shifts your focus forward. It's time to get moving and make tracks on developing your goals. It does see a time of significant opportunity is arriving that draws advancement into your life. A flurry of potential sets an enticing pace, it does bring heightened movement, a big reveal is that the basis of this exciting chapter. Create space for an upgrade, that restless feeling is guiding you towards an exciting adventure. There is magic ahead, you brew a potent elixir that enables you to dive into a refreshing chapter. There is a source of power arriving that eclipses your fears and doubts. Harnessing your tenacity, resilience, and confidence, you embark on a driven chapter that enables you to make some pretty substantial progress on achieving your vision. There is an opportunity to replenish and rejuvenate your spirit ahead.

Additionally, there is a secret being kept about you, but this person has no nefarious intent, it's quite the opposite. There is an attraction simmering below the surface that makes them stop in their tracks whenever they see you. They do hope to develop a closer situation yet feel shy about sharing their deeper feelings with you. They are looking for an opportune moment to get the ball rolling on exploring the possibilities further with you.

FEBRUARY WEEK FOUR

The Full Moon in Virgo occurs this week, this can create energy peaks that illuminate and draw clarity. It is a time of increasing potential, you are ready to expand your horizons, and this openness does bring a happy chapter that offers a variety of new possibilities. Information arrives that brings a much-appreciated boost, you discover events on the horizon lead to a richly creative and expressive journey. It does secure a time of advancing your vision. That lets you appreciate how far you have traveled as you plot a course towards developing your life. You are headed towards a positive chapter that gives you a nudge in the right direction. In fact, you're more than ready to create change in your life, so don't let doubt derail your progress. You are in the process of transforming your life and shall see a reflection returned that depicts the work and dedication you give. A flash of an idea may arrive suddenly and sweep you off to a new adventure. Your creativity is going to take you places, it brings new options that bring out your optimistic side. Surprise news calls your name soon. You are on a path to revolutionize your world. It does see a transition occurring that transforms your trajectory and brings new goals into focus. A vision you set your sights on does offer room to grow. A message arrives that sparks new potential. It does bring a good situation that is perfect for your talents. As this area makes an appearance in your life, you kick into high gear and shift your focus towards progressing your situation forward. It does launch a new and industrious chapter, it supercharges your dreams and gives you plenty to focus on in the section ahead.

MARCH ASTROLOGY

March 6 - Mercury Greatest Elongation.

The planet Mercury reaches its greatest elongation of 27.3 degrees from the Sun. If you would like to view Mercury, look for Mercury low in the eastern sky just before sunrise.

March 6 – Last Quarter Moon in Sagittarius.

This Moon phase occurs at 01.30 UTC. –

March 11 – Neptune in Conjunction with the Sun.

The planet Neptune in Conjunction with the Sun. This occurs at 00:00 UTC.

March 13 - New Moon in Pisces.

The New Moon creates space for a new chapter. This phase occurs at 10:21 UTC. This is an excellent time to observe galaxies and stars because there is no moonlight to obscure your view of the universe.

March 20 - Vernal Equinox.

The March equinox takes place at 09:37 UTC. There are equal amounts of day and night throughout the world.

March 21 – First Quarter Moon in Gemini.

This Moon phase occurs at 14.40 UTC.

March 26 - Venus Superior Conjunction.

The planet Venus at Superior Conjunction. This occurs at 06:00 UTC.

March 28 - Full Moon in Libra.

This Moon is on the opposite side of the Earth as the Sun and shall be fully illuminated. This phase occurs at 18:48 UTC. This full moon is known as the Full Worm Moon. Powerful energy lights a path forward. You can attract and manifest excellent results during the complete moon phase.

MARCH WEEK ONE

Mercury reaches greatest elongation this week. This can feel destabilizing, but in fact, it creates a useful change, you move away from destructive influences and create lifestyle changes that draw benefits. It's a productive time that arrives to put the polish on your talents. It does take your skills up a notch, you brush up on your abilities, and expand your expertise into new areas. Start with intentions and the value of planning the path ahead. Creating a realistic plan does mark the steppingstones toward success. A burden that you have been carrying is lifted soon. There is a significant life improvement coming, it draws a dynamic time that is active and productive. Your confidence is on the rise, so don't let old insecurities hold you back. Devoting your energy towards a passion project does advance a venture that is close to your heart. It has you feeling inspired; this enthusiasm permeates your surroundings. A moment of transformation shifts your focus forward. It's time to get moving and make tracks on developing your goals. It does see a time of significant opportunity is arriving that draws advancement into your life. A flurry of potential sets an enticing pace, it does bring heightened movement, a big reveal is that the basis of this exciting chapter. Create space for an upgrade, that restless feeling is guiding you towards an exciting adventure. There is magic ahead, you brew a potent elixir that enables you to dive into a refreshing chapter. There is a source of power arriving that eclipses your fears and doubts. Harnessing your tenacity, resilience, and confidence, you embark on a driven chapter that enables you to make some pretty substantial progress on achieving your vision. There is an opportunity to replenish and rejuvenate your spirit ahead.

MARCH WEEK TWO

Neptune arrives in conjunction with the Sun this week. The Planet Neptune rules dreams and healing, while the Sun places a strong focus on self-development and improvement. This can usher in many changes. It does definitely grow your potential, the influence of the Sun draws an energizing power that bolsters your spirit. You're ready to turn over a new page, something is refreshing in the wind that is going to flow into your world soon. Start with intentions and the value of planning the path ahead. Creating a realistic plan does mark the steppingstones toward success. A burden that you have been carrying is lifted soon. There is a significant life improvement coming, it draws a dynamic time that is active and productive. Your confidence is on the rise, so don't let old insecurities hold you back. Devoting your energy towards a passion project does advance a venture that is close to your heart. It has you feeling inspired; this enthusiasm permeates your surroundings.

Additionally, the New Moon in Pisces this week sees some unusual changes are occurring as you discover new information, someone has been keeping something back, it does lift the lid on a new chapter. Hearing what this person has to say may come as a shock, but the outcome is sweet as sugar. This is someone supportive, they energize you, and it does bring an enchanting time of sharing thoughts and ideas. This person is ready to share their secrets with you, it does bring a golden moment that is given from the heart. It primarily draws an environment that is healing and abundant. Intentions are set at this time, go far, and you see a goal reach fruition the next full moon.

Ostara, the Spring Equinox, takes center stage this week. It's all about facing the sun again after the long winter and starting new plans and goals, you can harvest later in the year. You weave a spell of manifestation; it draws exciting new options into your life. You may have been navigating a time of uncertainty, this could see some long-held beliefs challenged as you break through and enter a bold phase of evolving your situation. Removing limiting beliefs does create a clean slate potential. Something arrives that rings true, this message speaks to your heart, it reverberates loud and clear that change is imminent. You enter an active phase that is highly productive and enables you to make progress on many areas that have been on the backburner for a while. You are in your full glory, it does ignite a time that is busy and expressive. There is an insatiable yearning to reach for more, your creativity is heightened, and this boosts your spirit and helps you define goals to chase over the coming months. It is a time of luck and expansion.

Furthermore, impeccable timing aligns your path towards an abundant chapter. The aim is remarkable, you win over someone who becomes an essential focus in your life. This person has a great deal of discipline and integrity, they offer wisdom, support, and kinship. You have learned some crucial lessons and now can step into a phase that helps you attract the right people at the right time. Working with this person lets you keep your vision on track.

MARCH WEEK FOUR

Venus sashays into your life this week, it is a time that draws fulfillment; it provides you with outlets to nurture bonds that hold meaning. It enriches your life and brings rejuvenating pastimes and activities into your life. It's an ideal opportunity to become more social and tap into a more full arena of community options. Your creative aspect is also on the rise, it is a time of breakthroughs, inspiration, and adventure. An active, busy time has you feeling productive, capable, and this draws advancement. It does create a forward focus, movement towards your vision nourishes your life with exciting results.

Additionally, the Full Moon in Libra highlights a time of increasing opportunity, a major makeover is coming. The exciting news is coming soon, it is a game-changing chapter. Things are heading towards an upswing. Fortune shines upon your life, it does take you to a time of heightened opportunities. Being highly creative, you discover innovative solutions, enabling advancement to flow into your world. Opening the door towards a chapter which is exciting and active do draw joy. Moving towards growing your talents, does see you make progress on the path towards developing your vision. You are ready to turn the dial towards growth, you can make headway on your goals through the combination of proactive measures, and planning your strategy. Seeing the trajectory necessary gives you a leg up towards achieving the growth you are seeking. You soon can climb the ladder towards success; in fact, your best qualities enter into the spotlight when you receive an offer that feels meant to be. This potential arrives soon, it breathes new life into your world.

APRIL ASTROLOGY

April 4 – Last Quarter Moon in Capricorn.

This Moon phase occurs at 10.02 UTC.

April 12 - New Moon in Aries.

The New Moon phase occurs at 2:31 UTC. This is an excellent time to observe galaxies and stars because there is no moonlight visible.

April 19 – Mercury at Superior Conjunction.

The planet Mercury at Superior Conjunction. This occurs at 02:00 UTC.

April 20 – First Quarter Moon in Leo.

This Moon phase occurs at 06.59 UTC.

April 22, 23 - Lyrids Meteor Shower.

The Lyrids meteor shower runs each year from April 16-25. This meteor shower peaks on the night of the 22nd and the morning of the 23rd. These meteors can produce bright dust trails that last for several seconds.

April 27 - Full Moon in Scorpio, Supermoon.

The Moon is on the opposite side of the Earth as the Sun and will be completely illuminated. Full Pink Moon. It's the first of three supermoons for 2021. This occurs at 03:31 UTC. The Moon will be at its closest approach to the Earth and may look slightly larger and brighter than usual. Powerful energy lights a path forward. You can attract and manifest excellent results during the full moon phase.

April 30 – Uranus in Conjunction with the Sun.

The planet Uranus in Conjunction with the Sun. This occurs at 21:00 UTC.

Momentum builds quickly when a new possibility drives the path towards a growth-driven chapter. You think on your feet and make a value judgment about an area that tempts you forward. It does see your potential soaring upward, it is a golden time to chase your higher vision. A plan that is set into motion does bring an upgrade you can appreciate. Being open to the unexpected broadens your horizons in the most remarkable fashion. You enter a dynamic time that is vital and energized. It does let you harness the power of your inspiration to create essential lifestyle changes that draw a happy vibe. It includes foundations for a chapter that drives home the abundance with surrounds your world. There is advancement coming that brings palpable recognition of the hard work you have put into your situation. An exciting offer makes itself known. Revolutionary changes are transitioning you forward. This lets you break free from the past, you light up a new pathway, it revamps your potential. Restoration of spirit motivates you to chase your dreams.

Additionally, it is an ideal time to explore the broader world of potential which surrounds your social life. There is a gathering ahead that links up to new friends, it really sparks new options into your world that energizes and rejuvenates. New beginnings ahead support your efforts to advance your situation. Setting goals for where you'd like to be in several months is beneficial.

The New Moon in Aries packs a nugget of wisdom. It is a time of excitement, inspiration, and creativity, it does have you dreaming big about future goals. The world is your oyster, there are many options at your disposal. You discover a lot is going on in your life, the pace and rhythm are increasing, this helps you reach a higher trajectory. Something you have been hoping for is on offer. An offer arrives to improve your situation. A piece of the puzzle is revealed, you discover you can make tracks and achieve traction on a long-awaited goal. It takes you on a path of adventure and lets you transition to a working environment that holds great promise. Good news is arriving, you can clear the decks and create foundations that offer you room to grow and prosper. You release limitations, areas that had felt blocked suddenly shift forward, bringing with it an exciting world of possibilities. You fly high and harness inspiration and enthusiasm to stunning effect. It represents the completion of a long process of searching for the right role for your talents. This serves a new adventure, you harness a pioneering spirit and blaze towards a trail that calls your name. You embark on a journey that speaks to your mind and to your heart. The sun is behind you, this represents support is within reach, you are ready to plot a course and navigate around any hurdles that crop up. It is a time of significant accomplishment and change, you transition and evolves your situation on several levels.

Mercury at Superior conjunction this week sees you reaching a crossroads, it can be challenging to know the path ahead when you face a dissecting road. A refreshing twist ignites your creativity. It brings an offer; you reveal an opportunity that lets you contemplate the path ahead and make tracks towards a chapter of growth and stability. It brings a valuable sense of belonging, news arrives to point you in a fresh direction. It is a favorable time to upgrade your dreams and go after what you truly seek and desire in your life. It denotes a fortunate chapter where you can make tracks on achieving your loftiest goals. In fact, you blaze towards a path that sparks your passion, this venture arrives to kick start a new chapter of potential. It's an excellent time for setting intentions. It does bring new goals to light, your perception is broadening and taking you on a brighter scope of what is possible in your world. You can set your sights on a significant achievement and feel supported in your quest to better your environment. Advancement is looming, an offer is imminent that provides you with a valuable sign to head towards. It does bring a lift to your career goals, in reaching for a substantial path, you embrace a happy chapter. It brings a new area that sees you learning and growing your skills. It does create a shift forward that ushers in a new phase of life. It brings a reboot, you feel rejuvenated and can embrace a luxurious outcome. You set your life ablaze with new options, it is just the avenue you were seeking. It brings a valuable shift forward.

APRIL WEEK FOUR

The Full Moon in Scorpio is a Supermoon that sweeps into your life to draw closure, healing, and this transitions you forward. There is a mystery surrounding you; something you seek is going to be revealed. You may not necessarily see a clear path ahead, however, feeling lost or unsettled is not necessarily a bad thing, it draws new and more significant options into your life, it helps you find a path that calls to your heart. There is some rapid expansion coming, it draws change, and releases any holdups that have stifled your potential. If you are waiting for a sign, you can embrace the chapter ahead as it speaks to your intuition; it provides you with a guiding sense that you can understand and utilize to your advantage. Emotional intelligence is rising, your creativity is headed to the stratosphere, it brings an innovative time that sees you reawaken to the possibilities surrounding your life. It is a time that offers you room to reboot and rejuvenate. It does help you take stock of your goals, and you begin to envision plans you can put into place over the coming chapter. It creates space for you to release outworn energy, and this clears the slate for new inspiration to arrive. It does see you riding a roller coaster wave of emotions when unanticipated options strike a note, it gives you a broader sense of what can be achieved. Creating space for something new does soon achieve the results you have been seeking. Inspiration strikes, an original path begins to open, it takes a little time to come together, you dip your toes into the brew of manifestation and enjoy thinking about the possibilities.

MAY ASTROLOGY

May 3 – Last Quarter Moon in Aquarius.

This Moon phase occurs at 17.50 UTC.

May 6, 7 - Eta Aquarids Meteor Shower.

The Eta Aquarids meteor shower runs annually from April 19 to May 28. It peaks this year on the night of May 6 and the morning of May 7.

May 11 - New Moon in Taurus.

This phase occurs at 19:00 UTC. The new moon phase is a brilliant time to observe galaxies and stars because there is no moonlight visible.

March 17 - Mercury Greatest Eastern Elongation.

The planet Mercury reaches its greatest eastern elongation of 22 degrees from the Sun. If you would like to view Mercury, look for the Mercury low in the sky just after sunset. This planetary phase occurs at 06.00 UTC.

May 19 – First Quarter Moon in Virgo.

This Moon phase occurs at 19.13 UTC.

May 26 - Full Moon in Sagittarius, Supermoon.

This phase occurs at 11:14 UTC. Full Flower Moon. It's the second of three supermoons for 2021. The Moon will be at its closest approach to the Earth and may look slightly larger and brighter than usual. Powerful energy lights a path forward. You can attract and manifest excellent results during the full moon phase.

May 26 – Total Lunar Eclipse in Sagittarius.

A total lunar eclipse occurs when the Moon passes completely through the Earth's dark shadow or umbra. During this type of eclipse, the Moon gradually gets more mysterious and then take on a rusty or blood red color. This eclipse occurs at 11:19 UTC.

May 29 – Mercury Retrograde begins in Gemini.

During a retrograde period, it isn't the right time to move forward in any practical venture. Be prepared for misunderstandings and miscommunications to be prevalent.

The foundations of your life stabilize, it does bring a grounded and practical time. Planning the moves ahead draws a productive and lively chapter. You nail your objectives and envision new areas to develop. Climbing the ladder of success does see a sense of accomplishment shining in your life. You immerse yourself in a stimulating environment. The landscape is vibrant, it's full steam ahead, things are on the move you finally get to see the bigger picture and appreciate the results obtained from the work you have put into your situation. There is an emphasis on improving your bottom line, your schedule is filling fast, it is the perfect environment from which to map a solid plan for future goals. This is especially key as it lights a path towards success. You pour your energy into an area that rewards you with magic and manifestation, it's an auspicious time that opens a gateway towards improving your situation. There is a focus on business-related matters, constructive dialogues with colleagues give you a sense of the potential possible. You blaze towards achieving your vision, information arrives that motivates growth and progression. Your skills flex their muscles, you were given room to grow and take your abilities to a new level. It facilitates a vibrant time of advancing your situation. As you reveal Golden options, there is news imminent that can be used to take your vision further. A feasibility study may be required to determine whether or not this area is the correct fit for your career path. There is a cluster of activity coming into your life, which leaves you feeling exuberant. This expansiveness does spark your curiosity; it has you sorting through several options and launching towards a phase of growth and good fortune.

MAY WEEK TWO

The Taurus New Moon this week says that it's out with the old and in with the new. This is a time where your feelings may fluctuate if you feel sensitive or triggered by any issues, take time to focus on nurturing your spirit. You have strong intuitive abilities which enable you to read the fine print in people's eyes, tuning in on their body language helps you ascertain their true intentions. There is going to be an unprecedented surge of social opportunities, this takes you to a time that is busy and enjoyable. Your skills of organization help create the right environment for things to run smoothly. You can expect a myriad of offers to flow into your world soon. Life is going to be active; looking ahead, you feel encouraged to size the movement and embrace the change which seeks to flow into your world. You are guided towards a destination that holds promise; this suggests some significant happenings sweep you towards your dreams. You find your situation begins to resonate with harmony, balance, and emotional fulfillment. You become involved with one who speaks to your heart. You overcome hurdles by staying true to your spirit. Sometimes the path isn't as clear-cut as you would expect to be. This guides you to take the initiative and dig deeper into where you want to be and how you are going to get there. Channeling your energy correctly does clear the fog and takes you to a higher road less traveled. It also sees a fresh wave of energy arrive to support your dreams. You have accomplished a great deal, and as you reflect on the changes which swirl around your life, you gain valuable insight into the path ahead.

MAY WEEK THREE

Mercury reaches greatest elongation from the Sun this week. This reveals that you are positioned to advance your life. It centers on developing a situation that brings magical potential into your world. You are right to be curious and investigate leads, expanding your horizons is going to carry a new mission. Fortune and good luck guide a path forward, and unexpected opportunity does spark your interest; you pursue this avenue, and it does bring an active phase of creativity and expansion. Exciting developments ahead draw a triumphant chapter where you can obtain steady advancement. It does clear out areas that didn't reach fruition, organization, streamlining, discarding regions that are no longer useful, does heighten the potential possible. It takes you towards a time that offers advancement and progression. You spread your wings and embrace a time of adventure and discovery that is in sync with your hopes and dreams. You draw stability into your life; it re-fuels your emotional tank. It does bring a chapter that features new people, a friendship springs to life. This expansion in your social circle is welcome, it lets you turn a corner and embrace an active, lively environment. Discussions ahead draw bonding, it is a happy shift that epitomizes your willingness to open your life to new experiences. You write a wave of hope for energy and discover a path that boosts your morale. It is a winning time where you reawakened to the abundance that seeks to tend you forward. Circulating in a social landscape, you strut your stuff and emerge as a butterfly, metamorphosis complete, you are ready to soar high and experience a new outlook that is in rhythm with the person you are becoming.

MAY WEEK FOUR

This week delivers a plethora of cosmic activity. There is a full moon in Sagittarius, which is also a super moon, and on the same night, a total lunar eclipse. This is a triple magnifying event. But be warned, three days later Mercury retrograde begins in Gemini, this is the mule kick that may just knock you sideways if you're not aware, that it is coming. So what does all this mean for your life? The triple combo event on Wednesday brings changes that spark a new path. You should seek new experiences and keep open to learning areas that trigger your intuition. You nurture a sense of magic and adventure; it sets the stage for a happy chapter, your insights and ideas draw an abundant journey of exciting possibilities. Fortune shines upon your life. The wheels turning in your favor. A theme that resonates abundance puts a strong focus on achieving your dreams. You begin to see tangible results from work undertaken on developing your situation. This does wonder for your sense of confidence, it gives you a push in the right direction, you begin to feel you can achieve a higher result. Progress is dynamic and fluid, it's guiding you towards, a chapter which energizes and inspires. This takes you to a turning point, once at the crossroads, you choose a path wisely. You benefit from an active chapter ahead, in a whirlwind of activity, you build a basis which is durable and stable. This increases happiness and harmony in your life. This becomes the platform from which to sow long-term goals. Life blossoms and you can appreciate that the efforts undertaken have been well worthwhile to reach this new level.

JUNE ASTROLOGY

June 2 – Last Quarter Moon in Pisces.

This Moon phase occurs at 07.24 UTC.

June 10 - New Moon in Gemini.

This moon phase occurs at 10:53 UTC. This is an excellent time to observe galaxies and stars because there is little moonlight to obstruct your view.

June 10 – Annual Solar Eclipse.

An annular solar eclipse occurs when the Moon is too far away from the Earth to completely cover the Sun, it results in a ring of light around the dark Moon. The Sun's corona isn't visible during an annular eclipse. This solar eclipse is visible in eastern Russia, the Arctic Ocean, western Greenland, and Canada. A partial eclipse will be visible in the northeastern United States, Europe, and most of Russia. This eclipse occurs at 10.42 UTC.

June 11 – Mercury at Inferior Conjunction.

The planet Mercury at Inferior Conjunction. This occurs at 01:00 UTC.

June 18 – First Quarter Moon in Libra.

This Moon phase occurs at 03.54 UTC.

June 21 - June Solstice.

The June solstice occurs at 03:32 UTC. The North Pole will be tilted toward the Sun, which, having reached its northernmost position in the sky, will be over the Tropic of Cancer at 23.44 degrees north latitude. This heralds the first day of summer (summer solstice) in the Northern Hemisphere, the summer solstice is considered one of the most important times of the year for many traditional cultures.

June 22 – Mercury Retrograde ends in Gemini.

You can now move forward with any delayed plans that you have been putting off due to the Mercury Retrograde phase. Relationships should soon improve as miscommunications are overcome

June 24 - Full Moon in Capricorn, Supermoon.

The Moons will be completely illuminated. This moon phase occurs at 18:40 UTC. Full Strawberry Moon. This is the last of three supermoons for 2021. The Moon will be at its closest approach to the Earth and may look slightly larger and brighter than usual. Powerful energy lights a path forward. You can attract and manifest excellent results during the full moon phase.

You have the stamina to stay on top of things during this Mercury Retrograde phase. You may face some current hurdles soon, the path ahead is tricky to negotiate. You have met this environment before, and you do your best to deal with the issues as they arrive. The hidden information is revealed, there is a secret that is being kept from you, the telling of this mysterious footnote does draw clarity into a past chapter. It sparks a path of healing, and it emphasizes dealing with your emotions. You go within and dig deep to reveal what is below the surface and currently hidden from view. Understanding the complexities involved does strengthen your resolve to improve your situation. It improves your foundation and creates space to plant the seeds for future growth. You remove the obstacles that have slowed you down, seeing the pattern does help you create changes that let you shift towards a new chapter. You sweep away all that no longer serves your purpose, this draws fresh energy and heightened potential into your life. Understanding the bumpy patches of the past does let you navigate around the hurdles of the future, you take the lessons learned, and explore innovative ways to improve your life and revitalize your spirit. It is a time of going within, it does suggest insight and clarity will help rejuvenate and rebalance your mood. It's an appropriate time to pause and reflect, a new flow of abundance is ready to emerge, drawing, stabilizing energy does prepare your soul, it clears any energy blocks and reestablishes a foundation from which to grow your world. It is marked by an ending or transition, you soon embark on a new chapter of potential. A time of reflection removes outworn and unresolved sensitivities that limit progress.

JUNE WEEK TWO

The New Moon in Gemini combines with an annular solar eclipse, this sees change ahead for you. You face a crossroads, and there is a decision arriving soon, which is a gateway to a happy chapter. Once you cross through this portal, you discover an option that burns brightly. It is a time of dramatic potential, releasing areas that no longer serve your higher purpose, sees an influx of potential flow into your world. You open your heart to a variety of options, this gives your creative life expression, it enables you to partner your talents with your aspirations. As you unfurl a new chapter of potential, you plant seeds that blossom serendipitously over the upcoming episode. You are doing the right thing by exploring a variety of options. This is a crucial phase for you, you want to make sure you choose the right direction. As you are currently transitioning towards a new chapter, you find yourself at a crossroads, being indecisive is natural, but you can trust your intuition to guide you correctly towards developing your future. Luck is coming your way, this sets the stage for opportunities to socialize, it provides you with a valuable respite from the demands which have drained your energy. It's also a favorable time for reunions, meeting up with old friends and colleagues shine a light on meaningful conversations, it takes you to a time of lively discussions and collaboration. You find people who are a curious mix, and this captures your interest. This is an exciting time, which resonates with creative expression and the seeking of fulfillment. Finding the right balance between stability and expansion will enable you to move forward in a grounded manner.

The June 21st Solstice at weeks end is an ideal time to reflect on your goals. There is a new chapter coming which beckons and calls your name. You enter an energizing phase, which enables you to create essential changes. This resets your potential, it offers you a path towards developing an area that inspires your mind and cultivates your creative side. Fantastic opportunities on the horizons let you spread your wings, it is a catalyst for change, faith walks beside you, the path ahead glimmers with exciting possibilities. Your imagination is fueled by creative thinking. As you release the heaviness that has impeded growth, you are ready to launch towards an exciting journey forward. It does bring a gateway, a breakthrough moment, your passions come into focus, you take the first exciting steps ahead and can feel the shift that propels you towards chasing your dreams. It's a chance to liberate your mind from doubt and anxiety. You may discover your vision is changing, and as your awareness shifts, you get a new viewpoint that does help you get in touch with your goals. The intentions planted during this phase a given a chance to grow over the coming months. You can map a strategic plan and plot a course forward towards improving your circumstances. Clearing away any old vibrations enables you to draw new opportunities into your world. The soul-searching you have been doing is not in vain; you going to become a lighter and more inspired to seek the attainment of dreams. This kicks off an exceptional time, you put yourself out of all of your comfort zones when you make a move that surprises even yourself.

JUNE WEEK FOUR

Mercury Retrograde ends this week. Memories of the past may be tugging on your awareness to encourage you to pause and reflect. This creates space to resolve any residual feelings that may be clinging to your spirit and limiting growth. You might be doing some soul-searching or deep emotional processing, questioning the path ahead. If you are dealing with complex emotions, taking time to create space to nurture wellness and healing will help you work through unresolved feelings. Processing your thoughts will help you get your bearings; it does bring a time that embarks on a path where you can deal with your stuff. A new flow of energy in the form of a brilliant idea arrives to let you create headway on achieving your vision. It opens a gateway towards a prosperous cycle, a new venture brings exciting prospects. It does see you starting an original path, the lifestyle alterations tweak your situation. If you have found you have drifted off course and are unsure of the route ahead, you soon discover a direction you can head toward. Some calculated tweaks bring you in alignment with your vision. It connects you to a happy chapter that leaves you feeling rejuvenated and inspired. It is a compelling time where you create your own opportunities and go for gold. It takes courage to push back the barriers, but the dividends are worthwhile. You soon cast a light on something hidden, secrets are revealed, it gives you a glimpse of the past situation. It does bring a great time for resolving emotions and addressing healing in the past. It marks the ending of a chapter; this clears the way to transition you to a path that is in alignment with the person you are becoming. Change is surrounding you; you can discover an option that speaks to your heart.

JULY ASTROLOGY

July 1 – Last Quarter Moon in Aries.

This Moon phase occurs at 21.11 UTC.

July 4 - Mercury at Greatest Western Elongation.

The planet Mercury reaches greatest western elongation of 20.6 degrees from the Sun. If you would like to view Mercury, look for Mercury low in the eastern sky just before sunrise. This planetary phase occurs at 20.00 UTC.

July 10 - New Moon in Cancer.

The New Moon draws rebirth and new energy. This moon phase occurs at 01:17 UTC. This is an excellent time to observe galaxies and stars because there is no moonlight visible.

July 17 – First Quarter Moon in Libra.

This Moon phase occurs at 10.11 UTC.

July 24 - Full Moon in Aquarius.

The Moon is located on the opposite side of the Earth as the Sun and will be fully illuminated. This phase occurs at 02:37 UTC. This full moon is known as Full Buck Moon. Powerful energy lights a path forward. You can attract and manifest excellent results during the complete moon phase.

July 28, 29 - Delta Aquarids Meteor Shower.

The Delta Aquarids meteor shower peaks on the night of July 28 and the morning of July 29. The first quarter moon may block many of the fainter meteors this year. You should still be able to view some brighter ones. Best views should occur after midnight. Meteors radiate from the constellation Aquarius but may appear anywhere in the sky.

July 31 – Last Quarter Moon in Taurus.

This Moon phase occurs at 13.16 UTC.

Mercury at Greatest elongation this week brings unique vibrations. Something arrives, which is a culmination of a project, it is a matter involving business, it appears that you are now ready to reach for more. You are becoming stronger, braver, and more resilient. It does position you correctly to launch your potential forward. Setting your intentions lights the fire in your belly. The sparks of your inspiration do take you to a chapter where you can expand your horizons and obtain growth. It does bring new people into your life, and an adventure you can embrace. It recognizes your efforts drawing success. It is a time where you gain traction on your goals and can focus on achieving a stellar result. A creative project takes off. There is a choice to be made ahead that takes you to a happy chapter. It does suggest you find a community of people who love and support you. It's only the beginning as more of these kindred folk will pour into your life. You are attracting the ones you need in your world. It does bring engaging banter and lively conversations. Rapid expansion does shift your focus forward, it transforms the potential surrounding you. It is a time that lets you get out and circulate, as you embrace a social chapter, you discover a crew that puts the spotlight on abundance. These kindred spirits understand your quirks and sense of humor. It does bring quality time with friends, it may see you attending community events, meetups, and travel is an option. It sparks a lively and social time. Listening to the wisdom shared does bring a thoughtful path to explore. It is an ideal time to create space to nurture your world. Prioritizing wellness does draw dividends. Making sure you get the necessary self-care does cultivate an abundant frame of mind.

The New Moon in Cancer this week signifies a new beginning. News arrives, which initiates a wave of potential flowing into your world. It brings transformation, a breakthrough moment is imminent. It sets the stage for discovering the outlet that inspires your mind. You journey through a time of growing your world, it does reveal an area that spikes your curiosity. It is a time that refreshes and rejuvenates. Something exciting is coming; this brings a lively chapter. Developments ahead let the pieces of your puzzle fall into place, drawing abundance and joy. Tweaking your lifestyle is fast-tracking a successful outcome. It does bring a vast time for emotional awakening; it eclipses the past by anchoring your potential in a future-facing aspect. It brings new potential into your life that activates a highly creative zone, it releases unresolved baggage, and life flows forward, enticing you toward a bold beginning. Something unexpected unfolds that transforms your experience. It does see you building a stable foundation towards a progressive chapter. Your prospects are ready to improve, sharing your gifts with a broader audience does debut your talents in the most magnificent arena. You are unique and gifted, discovering the right path to expand your situation does see limitations being released, you create space to channel your energy in an environment that offers you room to develop your abilities. It does spark a wonderful time of self-expression and growth. It does bring a highly creative journey, you blend ideas and come up with a path that is off the beaten track. This new trajectory does give you broader options, it utilizes your gifts to full effect. As you lay the groundwork, you build a foundation that offers you room to grow and prosper.

An opportunity arrives that is a catalyst for change. It does debut a new area and inspires a mission that helps you smash through your barriers and leap into something you. It catapults you towards a phase of growth, it sees a breakthrough that sweeps in and demands your attention. It brings a path that inspires your mind, it does harness your abilities, expanding your horizons lets you hit your stride. You soon land in a creative environment; it does spotlight a path that tempts you forward. It brings a transition that sparks a journey of self-development. Unleashing your talents in a broader arena does bring significant change. You have an opportunity to radically revamp your trajectory and share your unique gifts with others. It does bring people into your life for a reason; a wellspring of abundance flows into your life. Exciting opportunities are ready to roll into your world, it does bring a theme that touches down on connecting with kindred spirits. You become involved in an area that offers a chance to network within your social circle. Forging new contacts does bring friendships to light; it sees you working beautifully with people of a similar mindset. It does bring a communal effort; a shared vision is at the basis of this chapter. Information is revealed that illuminates a new path. It does present an exciting opportunity to join forces with another. Taking the time to explore this offer does let you make an informed decision. It takes your ambitions further. It brings lovely potential into your world. There is an added responsibility coming, it advances your situation, and brings a dedicated avenue of growth that keeps you busy.

The magic of the Full Moon arrives to mend the scars of the past. Dealing with areas that linger on your awareness does create space for healing. Under this influence, you transform your potential, and this is instrumental in letting something new into your world. As you dismantle any blocks that hold you back, you become ready to embrace a new chapter. It does lead to change that ignites your imagination. Taking time to focus your energy and nurture your situation does build stable foundations. It brings a time of transformation around your life. You discover an area that inspires your soul, it places your vision forward and does end time of upheaval and uncertainty. You arrive at a gateway to a brighter future. Something that has been on your mind recently finally becomes clear. Pinpointing the next step to take does bring progress. You can refuel your emotional tanks, push back boundaries, and make strides on beginning that new chapter you have been contemplating. Suddenly, a visionary picture of what is possible arrives to inspire your soul. An unexpected opportunity has you exploring an entirely new path. This side avenue does represent the next phase of your journey. You create pioneering waves; it inspires your mind and gives you a taste of what's to come. It draws an energizing time of making moves that are pivotal to improving your situation. It brings a vivid and dynamic chapter that grows your skills and takes your abilities to a new level. It sees an outpouring of creativity is possible, an endeavor you focus on does take flight, you watch your ideas blossom.

AUGUST ASTROLOGY

August 1 – Mercury at Superior Conjunction.

The planet Mercury at Superior Conjunction. This planetary event occurs at 14:00 UTC.

August 2 - Saturn at Opposition.

The beautiful ringed planet Saturn will be at its nearest approach to Earth and will be illuminated by the Sun. This planetary event occurs at 05:00 UTC.

August 8 - New Moon in Leo.

This moon phase occurs at 13:50 UTC. This is an excellent time to observe galaxies and stars because there is no moonlight to obstruct the view. A new chapter awaits an open heart.

August 12, 13 - Perseids Meteor Shower.

The Perseids meteor shower runs each year from July 17 to August 24. It peaks this year on the night of August 12 and the morning of August 13. The Perseids meteor shower is usually excellent viewing as the meteors are so bright and numerous. The moon sets early in the evening, leaving dark skies for what could be a unique show. The best viewing is from after midnight.

August 15 – First Quarter Moon in Scorpio.

This Moon phase occurs at 15.20 UTC.

August 19 - Jupiter at Opposition.

The Giant planet Jupiter will be at its nearest approach to Earth and will be at it's brightest. This planetary event occurs at 23:00 UTC.

August 22 - Full Moon in Aquarius, Blue Moon.

The Full Moon draws clarity and illumination. This phase occurs at 12:02 UTC. Full Sturgeon Moon. This year it is also a blue moon. This event only happens on average once every 2.7 years, giving rise to the term, "once in a blue moon." There are three full moons in each season of the year. But as full moons occur every 29.53 days, occasionally a season contains 4 full moons. The additional full moon of the season is known as a blue moon.

August 30 – Last Quarter Moon in Gemini.

This Moon phase occurs at 07.13 UTC.

Saturn at opposition this week brings new information that makes quite an entrance in your life. It brings options that ignite your interest. This propels your situation forward; it also brings a transition and an ending. A new chapter is on offer, taking steps to change your environment does channel your adventurous side. You discover you can explore a path that offers you room to grow and prosper. You harness a sense of wanderlust in this refreshing landscape. It is a fruitful time to network; news arrives to tempt you out into a community environment. It does make for a chapter of entertaining and communing with kindred spirits. A collaborative approach opens the door to a team project. This endeavor connects you to like-minded trailblazers, it ignites an innovative chapter that is highly self-expressive. Focusing on this area soon opens a path forward that charms and inspires. It lights a way that lets you make headway on expanding your life. It is a bold time that creates change, it says that you can push back your boundaries and explore an adventure-driven chapter. Moving out of your comfort zone activates a path that inspires your mind. It draws a highly productive cycle that is active and social. There is a lot to be achieved by embracing the news ahead. This jump-starts a new chapter, it does spark an urge to move forward and create changes that lead to a noticeable improvement. It speaks of an opportunity that captures your interest. This is something that isn't on offer every day, it does take courage to move out of your comfort zone, but the rewards are worthwhile. You enter a chapter that highlights creativity and advancement, it ramps up the potential possible in your life.

AUGUST WEEK TWO

The New Moon in Leo this week begins a new chapter. You are headed towards the time of opportunity that could take your life in a new direction. It does outline a journey towards abundance as you light up areas that speak to your heart. Following your inspiration is the short ticket to success. Your dreams are within reach; taking a courageous step towards your vision, you discover you can advance your life. You are more resilient than you currently realize. You are in a phase of profound transformation, it does bring a joint venture that places you in the box seat to develop an innovative business idea. It is a time that brings research, being immersed in an area that captures your imagination does align you towards a stellar phase of growth. Your visionary ideas hold water, and you can set sail towards advancing your career path. It is a time of new beginnings and creating the changes necessary.

Information arrives that lights a path beyond your current situation. It allows you to build something tangible and see the results of your emotional investment. It brings foundations from which to grow your world. It is an opportunity to expand your life and reach for your vision. An invitation ahead connects you to a sensitive soul, this person plays a pivotal role in future events. It draws a sense of connection that is refreshing. You have a busy time ahead, news arrives to tempt you into a community environment. It brings a chapter that inspires and motivates you to connect with kindred spirits. Lively communications draw abundance, there is harmony in your surroundings. These talks sprinkle your life with excitement and possibility.

Your situation flows forward in due course, it does restore balance and gives you an idea about you are headed towards a stable foundation from which you can progress your vision. It does serve up new dreams, and while you may seek to micromanage a plan towards success, this is an instance where it is best to let things gently unfold over time. It does take a little while to come together, your perseverance is rewarded with a transition to a new chapter. Pursuing your goals does bring the result you are hoping for. A shift occurs over the coming months, it brings satisfaction for a journey that has reached completion. Your new path is the epitome of security, it fits the bill, and is the perfect venture to focus your energy on. It lands in your lap most unexpectedly. Some notable changes are coming, which give you new options to contemplate. It does favor a practical and productive approach, utilizing a methodical system, you discover you can progress your goals, increasing stability and security. You reveal some powerful options which enable you to set the bar higher, an idea that has been on the back burner for some time suddenly gets a push forward and can be developed. It gives you options that are in sync with your evolution, and this leads to an active phase where you can harness the power of creativity to good effect. This is a significant time where you can focus on improving your environment. It does see creative changes that draw new energy into your life. If you have found things have been stale recently, this is set to lift with a refreshing flow of potential, which shakes up the options in your world.

AUGUST WEEK FOUR

The Full Moon in Aquarius at the beginning of this week is also a rare blue moon. There are some lovely changes set to flow into your life. This week speaks of an influence emerging that captures the essence of dreams. Going after your vision draws abundance, a path comes into focus, bringing a sudden, unexpected benefit. It does help you create substantial progress, and this environment draws stability; it lets you pursue your goals and develop foundations that improve your home life. It brings a vital shift forward you can embrace. You are ready to open the door to a fresh start, a decision made soon creates change. Following your intuition, you enter a dynamic chapter that initiates transformation. It does see essential adjustments taking place, evaluating your goals, streamlining your situation; you can release areas that no longer hold relevance. This helps you blend your energy into a more powerful essence. It is a time of digging deep and harnessing your resilience and fortitude. There may be hurdles impeding your progress, but rather than giving up, you have the strength of spirit to reach beyond and achieve growth. You meet challenges and continue to strive to better your circumstances. An area that you have been working hard on developing does bring results. It is time to restore balance and create space to nurture your soul. Prioritizing your needs does support well-being, it creates the right foundation from which to grow your world. It brings efficiency to your efforts and re-energizes your life with new potential. Do make time to relax and replenish your tanks. An opportunity ahead will take your full attention.

September 7 - New Moon in Virgo.

The Moon is on the same side of the Earth as the Sun and will not be visible in the night sky. This phase occurs at 00:52 UTC. This is an excellent time to observe galaxies and stars because there is no moonlight visible.

September 13 – First Quarter Moon in Sagittarius.

This Moon phase occurs at 20.39 UTC.

September 14 - Neptune at Opposition.

The giant blue planet will be at its closest approach to Earth, and its face will be illuminated by the Sun. This event occurs at 08:00 UTC.

September 14 - Mercury at Greatest Eastern Elongation.

The planet Mercury reaches greatest eastern elongation of 23.8 degrees from the Sun. This event occurs at 04:00 UTC. This is the best time to view Mercury. Look for the planet low in the western sky just after sunset.

September 20 - Full Moon in Pisces.

The Moon is on the opposite side of the Earth as the Sun, and its face will be fully illuminated. This phase occurs at 23:55 UTC. Full Corn Moon. This moon is also known as the Harvest Moon. The Harvest Moon is the full moon that occurs closest to the September equinox each year.

September 22 - September Equinox.

The 2021 September equinox occurs at 19:21 UTC. The Sun shines directly on the equator, creating equal amounts of day and night throughout the world. This is also autumnal equinox in the northern hemisphere and is considered a significant zodiac event for many traditional cultures.

September 27 – Mercury Retrograde begins in Libra.

During a retrograde period, it isn't the right time to move forward in any practical venture. Be prepared for misunderstandings and miscommunications to be more prevalent.

September 29 – Last Quarter Moon in Cancer.

This Moon phase occurs at 01.57 UTC.

SEPTEMBER WEEK ONE

The past has been a time of coming of age, the learning of wisdom has planted seeds you now can harvest. A venture ahead gives you an environment that is ripe for progression, it brings a well-connected and dynamic person into your sphere. This lets you share your talents and collaborate with others. Combining resources draws tangible results. It opens the door to discussions that spark innovative ideas. It is an excellent time to explore new horizons. It brings a spontaneous time that captures a lovely sense of adventure. Moving out of your comfort, so does bring leads to investigate. You chase an opportunity that brings new friendships; it draws a vivid and dynamic chapter that resonates with excitement. Investing your energy into a journey that inspires your mind is a wise move indeed. You make significant headway towards achieving growth. Your sense of purpose holds you in good stead, it directs your energy towards achieving the results you seek. Some unusual changes are occurring that open a path that draws excitement. It ushers in a time of change and sparks inspiration, it becomes a big focus for you moving forward. A compelling avenue is revealed, it is a curious sideline for your creative energy. Focusing on this area does draw dividends. It is a path charged with new potential and leads to an inspiring project that you can channel your excess energy into. It is the ticket for a productive chapter that draws abundance and joy. You shift your focus towards this project, creating growth, and activating creative abilities to stunning effect. It does seem that you're able to reshape potential and revolutionize your environment.

Neptune at opposition occurs at the end of this week. Neptune rules your house of dreams and healing. There have been some issues that have not been helping your situation, digging deeper and getting to the heart of the matter does help you remove the deadwood. Additionally, a sign is coming that enables you to bridge the gap between your aspirations and desired results. Life soon contrasts with events that shakeup your environment. It does bring a path towards expansion. You are no stranger to barriers, and while things can feel like an uphill battle, there is the promise of new options that tempt you forward. You show the initiative and make waves, forging ahead towards your vision does clear away the cobwebs, it brings a project you can embrace. Things are turning full-circle. Staying true to your vision draws a path that soon enough, leaves you feeling excited and inspired about the possibilities. As your situation flows forward, you move towards developing your idea. It brings a portal, a new chapter, this transitions you to an environment that is ripe with potential. There are indications that hidden information comes to light that brings a happy realization. You are ready to make progress on your personal goals. It does plant the seeds which blossom over the coming months. You connect with a more social environment; it leads to developing a friendship that has the potential to flourish next year. The more you expand your life, the more you discover that life supports your goals by bringing you into the right landscape to progress your vision. You are set to experience a bold new beginning, it sets you on a new and surprising journey. It is a time of adventure and excitement, dynamic energy ramps up the potential in your world.

Change is on the agenda for you. This shifts your focus towards a new area, which is likely to play a pivotal role in furthering your situation. It's best to pay close attention to your intuition, as this activates abilities which help guide you towards the right area. Your unique talents seek expression in a tangible form, it is a beautiful way to foster your skills. It is a time that offers you fantastic opportunities to tempt you towards expanding those horizons. You discover it is smooth sailing, and as you steer your ship into uncharted waters, you can release any pressure or anxiety that binds your spirit up into tight knots. There is a role coming which is perfect for you, this arrives soon to lift your spirits, it does lead to a lucky chapter, which offers you room for progression. You may feel hesitant about accepting this offer at first, but as you discover newly found confidence, you get a strong sense of achievement as you overcome hesitancy and can appreciate how far you have traveled. New options are revealed soon, which lighten your load. Your sense of creativity is on the rise, and this leads to spontaneous outlets for self-expression, which draw a sense of rejuvenation into your spirit. Having an area to focus your restless energy on does bring you the chance to diversify your talents and explore a new realm of opportunities. Things are heading for an upswing for you soon. A unique opportunity is revealed, which leads to a path where you can explore options relating to career growth. Learning and development are at the crux of this energy. It does bring you to a phase where productivity is heightened, you can appreciate heightened security and even explore far-flung destinations as a future option.

SEPTEMBER WEEK FOUR

The Equinox this week speaks of a golden opportunity arriving to inspire your mind and shift your focus forward. This week reveals that you are a front runner for a chance that makes itself known soon. It does see you acting on your feet and grabbing the horse by the reigns, you know exactly what you hope to achieve by accepting this offer. It sees progression take center stage; it pares you up with a business opportunity that has you headed in the direction of growth. It gives you space to flex your muscles and dive into a productive chapter. There's a brighter world of opportunity ready to emerge. You enter a fascinating branch that treats you to a divine path of moving in alignment with your soul vision. It does shape your awareness as a time of transformation takes you towards building your dreams. Listening to your intuition guides the path ahead, it sweeps in a time of adventure and freedom. It is a fruitful time that brings a new friendship to light. There is a celebration ahead, the scene is social, discussions are animated and lively. It seems it is an excellent time to push back your barriers and step out in a community environment. Soon enough, you connect with others of a similar mindset. Exciting and inspired ideas spark a time of sharing thoughts and connecting with friends. It does indicate some hidden information will come to light during this time. Information reaches you shortly, which creates a breakthrough moment. It does see a situation turning in your favor. It is a time that brings unexpected news; this sees progress occurring soon afterward. A time of expansion brings a bountiful chapter.

OCTOBER ASTROLOGY

October 6 - New Moon in Libra.

The New Moon speaks of something new arriving in your world. This moon phase occurs at 11:05 UTC. This is an excellent time of the month to view galaxies and stars because there is no moonlight visible.

October 7 - Draconids Meteor Shower.

The Draconids meteor shower runs annually from October 6-10 and peaks this year on the night of the 7th.

October 8 – Mars in Conjunction with the Sun.

The planet Mars in Conjunction with the Sun. This occurs at 04:00 UTC.

October 9 – Mercury at Inferior Conjunction.

The planet Mercury at Inferior Conjunction. This planetary event occurs at 16:00 UTC.

October 13 – First Quarter Moon in Capricorn.

This Moon phase occurs at 03.25 UTC.

October 18 – Mercury Retrograde ends in Libra.

You can now move forward with any delayed plans that you have been putting off due to the Mercury Retrograde phase. Relationships should soon improve as miscommunications are overcome

October 20 - Full Moon in Aries.

The October full Moon is on the opposite side of the Earth as the Sun and will be fully illuminated. This phase occurs at 14:57 UTC. This full moon is known as the Hunters Moon. Powerful energy lights a path forward. You can attract and manifest excellent results during the complete moon phase.

October 21, 22 - Orionids Meteor Shower.

The Orionids meteor shower runs yearly from October 2 to November 7. Orionids meteor shower peaks this year on the night of October 21 and the morning of October 22.

October 25 - Mercury at Greatest Western Elongation.

The planet Mercury reaches greatest western elongation of 18.4 degrees from the Sun. Look for Mercury low in the eastern sky just before sunrise. This event occurs at 05:00 UTC.

October 28 – Last Quarter Moon in Leo.

This Moon phase occurs at 20.05 UTC.

October 29 - Venus Greatest Eastern Elongation.

The planet Venus reaches its greatest eastern elongation of 47 degrees from the Sun. This is the best time to view Venus. Look for the bright planet Venus in the western sky after sunset. This planetary phase occurs at 22.00 UTC.

OCTOBER WEEK ONE

News arrives that announces a significant option you can explore. It leads to a considerable time of growth and advancement. It has you thinking about fresh possibilities, it inspires your mind, optimism flows steadily around you. It does bring an option that lets you experience new sites and adventures. You spread your wings and can expand into a new area. This offers you room to progress your goals. The possibility of change is revealed when information arrives to provide you with a change of pace. It does pave the road ahead with magic and hope. An unexpected breakthrough offers a gateway to advance your situation. It is an exciting area that builds upon your current talents. Growing your skills is a surefire way to boot up your potential. Progress is achieved through your willingness to explore innovative pathways towards growth. Things ahead lineup to support your efforts as an exciting avenue opens that offers you room to progress a dream. It does bring abundance into your life; it allows you to tackle a long-held vision and get the ball rolling on achieving a stellar phase of progress. There is an avenue that helps you learn a new skill set; a specialist becomes involved with teaching or mentoring your situation. It does seem you benefit from a workshop or course ahead. This brings a time of higher learning and expansion; it does let you overcome barriers that impede progress. Taking the initiative clears the thorns from the path ahead.

Mars, in conjunction with the Sun this week, lets you pop the cork on the genie's bottle. It brings a sudden and dramatic entrance that requires your attention. Reevaluating the situation at hand, you can remove areas that are not helpful, and by getting rid of the deadwood, you break down your issues, and begin the process of rebuilding your life to create more stable foundations. Mars is an attention grabber, you discover there is something that requires decisive action. It is an excellent time for removing areas that limit progress. News arrives that transitions you to a phase of inspiration. It nurtures your abilities and creates an environment that evolves your skills. It does set the tone for a time of learning and acquiring skills that expand on your current talents. It opens a gateway that grows your potential, a variety of options arrive in quick succession, this spotlights a flurry of activity to tempt you forward. This is a valuable phase where significant change is possible. An opportunity comes that enables you to develop an area of interest. It's an extraordinary time that draws exceptional potential into your world. It charts a course forward that does offer new options, improvement flows into your world, it sets the ball in motion for an exciting chapter ahead. This sees a beneficial opportunity for money and growth arriving soon. It draws substance into your world. You embark on a new journey of discovery; it's a time that is exciting and harnesses an element of adventure as it does see you moving out of your comfort zone. It connects you to a broad array of potential, bringing moments that you can treasure. This propels you forward towards fantastic change, it's a time that glimmers with possibility.

OCTOBER WEEK THREE

This is a time that grows your world. Constraints are lifted when Mercury Retrograde ends this week; your situation expands outwardly. You can take advantage of an opportunity that offers you the chance to connect with your community environment. It does open your heart to new prospects, it brings a social environment that tempts you forward. Mingling with others draws the right kind of people into your world. A new friendship sparks to life, it brings a time that expands your world. A compelling path ahead draws an exciting new chapter. A refreshing viewpoint arriving to connect you with people and opportunities that link you up to the next part of your journey. Keeping an open mind does draw new experiences that tempt you forward. A whole new direction could unexpectedly arrive through your willingness to explore diverse possibilities. A changing scene is on the horizon that leaves you feeling inspired.

This Full Moon in Aries brings in sweeping changes, it does slow your situation down to some extent, yet this is seen as welcome, it gives you a chance to integrate these changes into your life. Focusing on emotional wellness, you process unresolved sensitivities. It is a sojourn through the past that heals and creates an environment that nurtures your soul. An option is arriving that bodes well for your life. It does draw answers to long-held questions. As you shift towards a new change of pace, you discover a direction that connects you to your dreams. It is a transition that reinvents and rejuvenates. A completely new perspective is possible, it brings a fresh start that lets you gain valuable traction on your vision.

An innovative decision marks a turning point. Something big is around the corner, you pitch your ideas and begin to tread a path towards your dreams. Taking that leap of faith does bring a great adventure. It draws a creative time that tempts you to grow your talents and believe in your abilities. Life expands as the world turns in your favor. Surprise news lights an exciting path forward. It is a brilliant time to develop a way that is in alignment with a long-term goal. It brings a transit, releasing doubts, does amp up the potential possible. It takes you to a time of building foundations that gently progress your situation. Taking proactive steps does advance your vision. Brick by brick, you build rock-solid foundations that blaze a trail towards an industrious chapter. You are ready to put your plans into action. Your perseverance and reliable determination draw dividends. You reach a peak and can make a move towards prosperity. It does bring a fresh start, you gain momentum on your vision, this sees the seeds that are planted blossoming into forwarding progress. Expect new possibilities that activate your creativity, it does see a path humming along that inspires your mind. Focusing on your goals becomes a priority, it brings options that enable you to build your talents. Your innate skills and abilities hold you in good stead, it reveals a path that brings joy. It does seem that you can trust your instincts on this one. Your prospects are about to become rosier, it lifts the lid on a chapter of potential that enables you to chase your dreams and expand your world.

NOVEMBER ASTROLOGY

November 4 - New Moon in Scorpio.

The New Moon brings a clean chapter of potential. This phase occurs at 21:15 UTC. This is an excellent time to view the stars because there is no moonlight visible.

November 5 - Uranus at Opposition.

The blue-green planet will be at its closest approach to Earth, and its face will be fully illuminated by the Sun. This event occurs at 00:00 UTC.

November 11 – First Quarter Moon in Aquarius.

This Moon phase occurs at 12.46 UTC.

November 12 - Taurids Meteor Shower.

The Taurids meteor shower runs yearly from September 7 to December 10. It peaks on the night of November 12.

November 17 - Partial Lunar Eclipse

A partial lunar eclipse occurs when the Moon passes through the Earth's partial shadow or penumbra, only a portion of it passes through the umbra. During this eclipse, part of the Moon darkens as it moves through the Earth's shadow. This partial lunar eclipse will be visible throughout most of eastern Russia, Japan, the Pacific Ocean, North America, Mexico, Central America, and parts of western South America.

November 17, 18 - Leonids Meteor Shower.

The Leonids meteor shower runs yearly from November 6-30. The Leonids meteor shower peaks this year on the night of the 17th and morning of the 18th.

November 19 - Full Moon in Taurus.

The Full Moon is on the opposite side of the Earth as the Sun and will appear fully illuminated. This phase occurs at 08:58 UTC. This full moon is known as Full Beaver Moon. Powerful energy lights a path forward. You can attract and manifest excellent results during the complete moon phase.

November 27 – Last Quarter Moon in Virgo.

This Moon phase occurs at 12.28 UTC.

November 29 – Mercury at Superior Conjunction.

The planet Mercury at Superior Conjunction. This planetary event occurs at 05:00 UTC.

The New Moon in Scorpio this week brings insight, clarity, and awareness. If life becomes complicated or chaotic in the weeks ahead, break tasks up into smaller, more manageable sections. Re-calibrating your energy, creating space to nurture your creativity, does draw balance into your world. Your ideas are ready to move towards progression. Taking steps to outline a plan forward does plot a chart towards success. You reach crossroads that requires decisive action. You may feel locked with doubt or anxiety if this time triggers resistance to change, nurture the faith that you are being guided towards the correct path. A transformation ahead draws abundance, it brings possibilities that activate creativity. It teams you with an area that offers you room to grow your dreams. As you navigate forward, a clear picture of what is possible emerges. A new phase of life is coming, it does bring decisions that can trigger sensitivities. Going within creates space to resolve uncertainty, integrating the changes ahead thoughtfully and gently does build solid foundations, and this is worth the investment of time involved. It brings a path that lets you venture further and explore possibilities that tempt you forward. You are set to experience a lovely boost, which helps rebalance flagging energy. If you have found yourself feeling depleted recently, you can expect power which offers you room to harness your creativity, it does bring beautiful changes that give you the freedom to focus on developing your talents. It does see incredible energy arrive, which leads to a breakthrough. In fact, it kicks off an active phase of growth, this provides you with generous opportunities to improve your situation. It is a time that reflects a journey to new heights.

NOVEMBER WEEK TWO

The Taurids meteor shower, which peaks on November 12th this year, see your potential shine brightly. This week speaks about an opportunity arriving soon that creates waves in your world; it connects you to the broader environment of potential, which is currently seeking to tempt you forward. It forms a basis of grounded energy from which you can forge a rock-solid foundation. This places you in the box seat to grow your vision and chase expansion. An opportunity is arriving that is meant specifically for you. It does open a path forward and strengthens your ability to improve your circumstances. It connects you with a project you can embrace.

Consequently, you are headed towards an ambitious and productive chapter. It is a great time to generate leads and obtain a progressive result. It is a lively and dynamic time ahead. New options arrive to bring you a lift. It does rejuvenate as it brings the power of renewal. The wheels of good fortune are turning in your favor; it sees you transition to a happy phase where you can see improvement over your situation. Shedding outworn areas resolves issues that limit or hold your progress back. A new perspective sweeps in exciting changes. Essential changes arrive that inspire you a great deal. You are headed to an extensive chapter that reinvents your potential. It is a time of change, life flows forward, stress levels are much lower, anxiety and doubt are released, and this makes way for a more balanced environment. Firm foundations are built that offer rejuvenation and happiness. A surprise makes itself known soon; it does give you a welcome boost. Finally, you feel things are moving forward.

NOVEMBER WEEK THREE

A Partial Lunar Eclipse on the 17th brings a landmark moment, it is a gateway toward a brighter future. This Lunar Eclipse speaks about a second chance, arriving soon for you. There is a beautiful symmetry in the chapter ahead, it draws healing and is quite therapeutic for your mood. It enriches your life by highlighting the path towards abundance. It brings a time where you make traction on achieving your objectives. Seeing tangible progress is exhilarating; it gives you the feedback necessary to continue this journey forward toward your vision. You are resilient and capable. It does hold you in good stead, helping you to power through the tough times, as this willingness to persevere takes you to greener pastures. In fact, a bountiful chapter ahead lets you get busy crafting your vision into a concrete path forward. A treasure trove of new options brings inspiration flowing into your world. It is a time that gently tempts you forward, there are options to develop your life and identity. You enter a self-expressive phase that harnesses the power of your creativity to create steady evolution. An area you discover becomes a priority, this gives you a clean slate of potential, it does bring a compelling path that inspires your mind, it marks a time where you can chase your dreams and relish the results possible. Expanding your life lets you focus on long-term prospects. A new cycle is coming, which allows the action you make to shape destiny. The more you put into improving your surroundings, the higher your results will be. A compelling path forward opens to guide you towards growth. Taking the road less traveled does prompt you to harness a sense of adventure and step out of your everyday routine.

This week speaks of change arriving soon. It does help to heal old areas that have clung to your spirit recently. There may be some forgiveness work needed to close a chapter that has been pulling on your heartstrings. It does bring a time of healing and closure to the forefront of your life. Doing this deep inner work creates space to rebalance your emotional outlook. It does bring a time of peace and contemplation. Releasing the substantial feelings soon sees new options flow into your world. If it feels like things are, no coincidence, this is your intuition guiding the path ahead. Fate mixes with destiny to arranges opportunities on your behalf. This infuses magic into your life and does help steer you onto the correct journey. This time speaks of a chance flowing into your life soon that brings a mystical element to light.; it does draw lifestyle changes, a new approach, and self-development take center stage. In fact, being mindful of the difficulties which surround your situation, help you make the most of untangling the issues, and this allows you to come up with innovative solutions which push your potential forwards. You discover you can strike gold, it draws a time of promise, it does see disruptive energy being released, you see your potential through entirely new eyes. It brings a breakthrough, where you can make progress on your vision. This takes you toward a journey of change, you soon get a clearer picture, it brings excitement and activity into your life. You turn a corner and headed towards growth, your willingness to persevere does draw dividends. Life begins to flow more easily, it brings exciting options to contemplate. A social situation emerges that is a salve that transitions you to a happy time.

DECEMBER ASTROLOGY

December 4 - New Moon in Sagittarius.

The New Moon brings a clean slate of potential. This moon phase occurs at 07:43 UTC. This is an excellent time to view galaxies and stars because there is no moonlight visible.

December 4 – Total Solar Eclipse.

A total solar eclipse occurs when the moon completely blocks the Sun, revealing the Sun's outer atmosphere, which is called the corona. The path of totality will, for this eclipse, be limited to Antarctica and the southern Atlantic Ocean. A partial eclipse will bee visible throughout much of South Africa.

December 11 – First Quarter Moon in Pisces.

This Moon phase occurs at 01.36 UTC.

December 13, 14,15 - Geminids Meteor Shower.

The Geminids meteor shower runs each year from December 7-17. The Geminids meteor showers peaks this year on the night of the 13th, 14th, and 15th. The nearly new moon this year will provide dark skies for an excellent show. Best viewing will be from a dim vista after midnight. Meteors will radiate from the constellation Gemini but can appear anywhere in the sky.

December 19 - Full Moon in Gemini.

The Full Moon illuminates and draws clarity. This moon phase occurs at 04:36 UTC. This full moon is known as the Cold Moon and the Moon Before Yule. Powerful energy lights a path forward. You can attract and manifest excellent results during the full moon phase.

December 21 - December Solstice.

The 2021 December solstice occurs at 15:59 UTC. The South Pole of the earth tilts toward the Sun, which, having reached its most southern place in the sky, is directly over the Tropic of Capricorn at 23.44 degrees south latitude. This December solstice also marks the first day of winter in the Northern Hemisphere.

December 21, 22 - Ursids Meteor Shower.

The Ursids meteor shower occurs each year from December 17 - 25. This meteor event peaks this year on the night of the 21st and morning of the 22nd.

December 27 – Last Quarter Moon in Libra.

This Moon phase occurs at 02.24 UTC.

DECEMBER WEEK ONE

December hits the right kind of positive note that you need in your life. Spending time with kindred folk is an act that nurtures your spirit. There are opportunities to mingle ahead that draw a sense of rejuvenation; it rejuvenates and renews your soul. If you have felt disconnected from your tribe, it is the perfect chance to tempt you out in a social environment. A whirlwind of activity brings the opportunity to get involved with lively discussions, and the sharing of thoughts and ideas, this change of pace is therapeutic for your soul. It brings momentous transition that shapes the course of your future life. It is a change-making time, it marks the entrance of a direction that nurtures your spirit. It does ignite the flames of reinvention. An area you nurture picks up speed, it does bring an elusive goal into focus. After all the confusion of the past, you can release outworn energy and embrace a clean slate of potential. This week speaks of change arriving soon to bless your life with a refreshing option. It does bring a goal that jump starts a new path. Knowing the correct direction to head toward does help you identify long-term objectives. It brings an enterprising chapter that is productive and lively. This fuels your creativity and brings expansion to the forefront of your life as you set your sights on a visionary goal. Your restless spirit is guiding you to step out of your comfort zone and mark a bold path toward your future. It does position you well to explore a new area.

There is a great deal on the horizon that arrives to tempt you towards expanding your world. It does launch your situation forward and, indeed, it's a beautiful turning point. Your curious mind is hungry for new adventures, staying open to people, resources, and pathways do feed your inspiration, it nurtures your spirit with exciting potential. Focusing on developing your talents attracts the right people into your life. You can embrace a time where your ideas gain traction, working to achieve your dreams does bring a vision to life. It is a steady progression that advances towards the goal post. Feeding your creativity is a fantastic way to nurture your abilities. It creates the energy to nourish your spirit. A curious enterprise arrives in your life. An ideal option takes shape, there is an unusual amount of activity coming to spark a productive phase. It does draw learning and growth. An innovative idea sparks a trail you can embrace. It galvanizes you to reach for your goals and set the bar on a lofty aspiration.

Furthermore, your social life becomes the focal point soon, it does have you diving into new depths, it deepens a bond which the one who captures your imagination. This draws harmony into your world. It does enable you to focus on connecting with your tribe and enjoying a more expansive social life. Surprising news flings open the door to a fresh start. It does have you landing softly in uncharted territory. Unexpected news arrives to light a path forward, it is harboring change and growth. Everything takes on a glow of inspiration, bringing a river of potential into your life.

DECEMBER WEEK THREE

News is coming within weeks, it does let you plot a course towards an exciting enterprise. It carries you forward towards building new foundations. As you secure your plan for the future, you discover that things are not as challenging to progress as first feared. You soon gain traction on your vision, drawing abundance into your world. It has you in touch with friends and contacts more than usual. This works to quicken the pace and bring opportunities to mingle into your life. It does let you make impressive progress, as you broaden your vision, you bring new energy into your world. It also suggests that a new element is going to play a crucial part in the developments ahead. You may have been feeling frustrated by lack of progress, there have been several issues which have been causing delays as they tend to gravitate potential to sidelines under a hectic life environment. However, you will begin to notice new options brighten your situation and put you in a phase of growth. It is a time of inspired possibilities, mixing and mingling with others, and spending time with a person who adds warmth and richness to your life. You soon reveal specific improvements are available, this becomes a phase which offers growth and stability. It ramps up potential, allowing you to debut a new chapter of potential. It's timely, fundamentally beneficial, and leaves you feeling happy to be sailing smoother waters. There is good news coming, this leads to some excitement in your house. You may be making some unique plans before long, and enjoy the bustle of recent activity.

DECEMBER WEEK FOUR

You are set to enter an optimistic chapter, it offers you options to circulate, this social growth does inspire a light-hearted environment. It has you spending time with friends and family. It does lead to a chapter of blessings and inspiration, which is highly motivating. You discover a robust and favorable transition is at hand, and this positions you towards a future that is abundant and adventurous. This is an essential chapter for you, which provides you with a transition forward. It does see potential is shining brightly for you. This helps you commit to a path that offers scope for improvement, and it does feel like something exceptional is unfolding in your world. As you reach for more, you enter a chapter blessed with harmony and joy. The coming new year ahead enables you to open the page on a new section. A vital change arrives; this rejuvenates, heals, and does have you thinking about the possibilities. It is a time that draws newness into your world, it brings options to advance your social life, this leads to new friendships. You are ready to embrace the future, it shines a light on mingling with others, leading to a fun time of expanding your horizons. Discovering a more connected crew of people to network with offers you room to grow your personal goals. You find that adventure is calling your name, this represents expanding your horizons and exploring new areas which are exciting and inspiring. Taking a momentous leap forward into the new year next week does lead to a journey of a lifetime. It's the perfect time to follow your heart and explore a path that speaks directly to your spirit.

Dear Stargazer,

I hope you have enjoyed planning your year with the stars utilizing Astrology and Zodiac influences. My zodiac star sign books are released each year, which detail a monthly list of astrological events, and a unique weekly (four weeks to a month) horoscope. You can find me on my Facebook page where you can get personal astrology or intuitive readings.

https://www.facebook.com/SiaSands

Instagram: SiaSands

See my full list of books here:

https://www.SiaSands.com

Leaving a review is welcomed and appreciated.

Many Blessings,

Sia Sands